French Artillery

French Artillery

by

Patrick Griffith

illustrated by

Emir Bukhari

Almark Publishing Co. Ltd., London

First Published 1976.

ISBN 85524 257 4

Distributed in the U. S. A. by
Squadron/Signal Publications Inc.,
3515E, Ten Mile Road,
Warren, Michigan 48091.

Printed in Great Britain by
Chapel River Press,
Weyhill Road, Andover,
Hampshire.
for the publishers, Almark Publishing Co. Ltd.
49 Malden Way, New Malden,
Surrey KT3 6EA, England.

Contents

Preface

In this book we shall be examining the tactical employment of French Napoleonic field artillery. Battery organisation and equipment will naturally have a place in the story, but so will the use of guns in larger units; the massing of up to 100 pieces for the decisive effort in a battle. It was this technique which contemporaries recognised as a significant new departure in the art of war, and which every nation in Europe copied from the French example.

The road towards the use of artillery in a decisive role was a long and tentative one. Many experiments were made, not all of them successful, before Friedland where Sénarmont was able to give his masterly set piece demonstration of what could be done. Artillerists had been talking of massed batteries since the days of Frederick the Great, but it required a variety of additional stimuli in the Revolutionary Wars before this could be translated into a reliable practical method. The increased mobility of French artillery under the Gribeauval system was extremely important, as was the related appearance of horse artillery. Of perhaps equal importance was the use of the divisional system, which encouraged commanders to look at their battles from a higher point of view and to mass reserves. In the case of the French the growth of the Imperial

Guard brought with it the creation of an *élite* artillery reserve upon which Napoleon came to rely for his victories.

As with all discussions of tactical practice there is a certain gap between what was done on the battlefield and what generals wrote in their drill manuals. In this case the use of massed artillery was scarcely mentioned officially before Napoleon was safely on St. Helena, and in fact it was not until 1809 that even an unofficial manual was written for the drills of a single battery. Previously the manuals had covered only the handling of individual guns. Yet ever since the early days of the revolution there had been attempts to mass batteries, and in every major battle after Eylau this was a central feature of tactics.

The gap between theory and practice meant that generals could follow their own inspiration as they pleased, and had no standardised drill to follow. They must nevertheless have discussed tactics among themselves, and there appears to have been a fair degree of unanimity between them as to how guns ought to be used. It is this unwritten code which we shall be describing in the following pages, as far as it can be disentangled from the mass of confusing evidence which has come down to us. Actual examples will be examined, as well as the practical operational habits which are mentioned in old soldiers' memoirs and privately printed textbooks. Out of all this it is quite obvious that Napoleon's field artillery was a major force to be reckoned with in all his later battles, and earned a formidable reputation among his foes. "Against that fellow", Blucher complained, "you need cannons, and lots of them".

Paddy Griffith
Sandhurst 1976

Field Artillery Equipment

It was the great artillery reformer Gribeauval who prepared the standardised equipment which was to serve the French so well in the Revolutionary and Napoleonic Wars. In the years after 1765 he completely redesigned ammunition, gun barrels, carriages, caissons (ammunition wagons), pontoons, and all the other vehicles and stores necessary for a mobile army. Despite considerable opposition at court he pushed through his reforms with the aim of giving the artillery two paramount qualities: standardisation and mobility.

The principle of standardisation is today accepted as essential to all military equipment, but in the eighteenth century it was revolutionary. Each manufacturing workshop would have its own foibles and peculiarities, even when it was supposedly working to a centralised pattern. For this reason the various types of equipment with an army would not have interchangeable parts, and repairs on campaign would be unnecessarily difficult. Spare parts could often not be fitted without alterations, and badly damaged vehicles could not easily be cannibalised to repair others. Gribeauval changed all that, and by the time of the Revolution it could be said that all French artillery workshops were producing identical components. Repairs in the field were considerably simplified.

It was the increased mobility of Gribeauval's guns, however, which gave the greatest delight to the gunners of the day. He cut back the lavish ornamentation which had previously encrusted gun barrels, and pared down their weight by about 45%.[1] At the same time he introduced several devices which allowed the guns to be manhandled with ease, thus bypassing the horse team for many movements in battle. In the first place he introduced a set of drag ropes (*"bricoles"*) and levers by which the gun crew could pull their piece easily in any direction. Secondly he used a split trail with a rounded base which did not stick in the ground when the gun was pulled backwards. Combined with this was the use of a long rope (known as a *"prolonge"*) which could be attached to the rear of the guncarriage at one end, and to the limber[2] at the other. With skilful use the gun could thus be fired while still attached to the horses, and then immediately set in motion without sticking on minor irregularities in the ground. This was particularly useful in fighting retreats, when the gun could be kept in action for the maximum time without needing the laborious process of lifting the trail onto the limber. French gunners would have been outraged, however, at any suggestion that their equipment was suitable only for retreats, and the *prolonge* was also very handy for rapid advances under fire.

Linked with the improved manhandling qualities of Gribeauval's guns was a greater robustness in his vehicles. Axles were made of iron instead of wood. Harnesses were made with the more efficient wooden poles rather than simply with ropes and straps. Wheels were increased in size to give better cross country performance, and when the heavier guns were moved from their trunnion position for firing to the travelling position the distribution of weight was more evenly spread between guncarriage and limber. All this made for greater reliability on campaign, although at the cost of a certain increase in the weight of the vehicles. While gun barrels became lighter the guncarriages and limbers were often slightly heavier than previous models, although when combined to make a complete team of gun and limber the net weight saving was still about 20%.

Not only did Gribeauval give the artillery greater mobility and reliability, but he also contrived to give it greater accuracy. In the first place his ammunition was made up into standardised cartridges which ensured that each shot was propelled by the same amount of powder as its predecessor. This meant that adjustments of aim could be more sensitive, as well as giving great advantages in handling the ammunition. Secondly the cannon balls were designed to fit more perfectly the bore of the gun, which reduced windage and again improved accuracy. Finally the aiming mechanism was much improved, with an adjustable backsight instead of a rudimentary notch on the barrel, and a delicate elevating screw instead of an unsophisticated wedge. All this allowed Gribeauval's followers to reply triumphantly to critics that although his guns were far lighter and handier than previous models, they could also shoot further and better.

As far as the field artillery was concerned Gribeauval used five basic types of gun. His four pounder was to be organically attached to infantry battalions, and it was this gun which was later to be widely used by the horse artillery. The eight pounder and 6.4" howitzer were to provide the backbone of the artillery concentrated at brigade level, while the twelve pounder and the 8" howitzer were for heavy support tasks in the general reserve. Thus the chain of command in the army's artillery was divided into three levels: battalion, brigade, and army. There was also the added complication that the vehicles were all driven by civilians who might not see it as their duty to manoeuvre under fire.

By the late Empire there had been several significant changes in this system, and an attempt had been made to streamline it. In the first place the use of divisional units had brought the transfer of most of the guns from brigade to divisional level, thus concentrating a powerful force under the divisional artillery commander. Secondly the battalion guns had finally been abandoned on 24th January, 1798, in order both to economise on the number of pieces, and to prevent the guns from slowing down the infantry. In the Italian campaign of the previous year thousands of Austrian infantry had been captured simply because they were left behind to protect their cumbersome battalion guns. Napoleon reintroduced this form of artillery for his raw troops in 1809, but it did not survive the Russian campaign.

Thirdly the civilian drivers were all militarised on 3rd January, 1800, as the "Artillery Train"; a much better arrangement which allowed the horse teams to come into battle, and ensured that ammunition wagons could always be kept within reach of the guns. Fourthly the horse artillery had been created; and finally a commission under Marmont reported in 1803 that several changes were required in the actual types of gun in service.

Marmont's new artillery became known as the "System of the Year XI", and differed from Gribeauval's in several important respects. Instead of a four pounder for light work and an eight pounder for medium work, Marmont wanted a single piece, a six pounder, to perform both tasks. There had been many complaints that the old four pounder was practically useless, especially with canister, and after the abolition of battalion guns there seemed little justification for retaining it. As for the eight pounder, it was thought to be too heavy for the job, although in the later battles of the

1 Characteristics of Artillery Vehicles

Type of Equipment	No of horses	No of shots in Coffret	Shots carried in each 4 horse caisson* Ball	Canister
12 pdr	6	9	48	20
8 pdr	6*	15	62	20
4 pdr	4*	18	100	50
6.4 inch Howitzer	4*	4	49	11
Forge	6	—	—	—
Caisson de Parc	4*	Carries spare parts, tools, or 12,000 infantry rounds and 1,000 flints.		

*Six horses instead of 4 in Guard of Horse Artillery Batteries

Napoleonic Wars the heavier metal would come to be more appreciated. Gribeauval's howitzers were also criticised as clumsy designs of limited range, and they were both to be replaced by a 5.7" model based on an Austrian pattern.

In addition to his recommendation to reduce the types of field piece from five to three Marmont also recognised certain other defects in Gribeauval's equipment. On campaign, for example, it had been found that the artillery could never have enough ammunition, so larger caissons were required. Also schemes were put forward to cure the tendency of Gribeauval's caissons to let in rain and spoil the cartridges. With the heavier bombardments of modern battles, too, French gunners had been inspired to use heavier charges than the guncarriages would bear, and even Gribeauval's robust designs had sometimes failed to stand the strain. Marmont thus wanted to design almost an entirely new set of equipment for the artillery.

The difficulty with this scheme was that it required lengthy experiments before the new models could be produced, and a massive manufacturing effort to replace all the older guns. In the event all this was only partially achieved, and instead of reducing the number of guns in circulation the effect was to add some new ones which took their place beside the old Gribeauval types and several foreign designs captured in the wars. There were, however, eventually enough of the new pieces available to equip the *Grande Armée* for the 1812 campaign, but the forces in Spain had to be content with a pure Gribeauval system.

Unfortunately the new equipment also showed signs of rushed design, and the six pounder guncarriage in particular had a tendency to shake itself to pieces on campaign. It was also deemed by many soldiers to be an ineffective compromise between the battering ability of the eight pounder and the mobility of the four. In 1818 it was abandoned and Gribeauval's system was officially revived.

Marmont's howitzers were also controversial, and both 6" and 8" variants were later added at different times, based on Prussian and Russian designs respectively. All these changes reflect the growing general tendency towards heavier guns in the later Empire, apart from the limited revival of battalion guns in 1809. Even then the four pounders would be the first to be abandoned in difficult conditions, and for the 1814 campaign there were very few of them with the army.

Apart from their use as battalion artillery the light pieces were also used by the horse artillery, although even here the tendency was towards heavier guns as the wars went on. This arm was created in France in 1792 with the aim of taking the principle of mobility as far as it would go. Every gunner had his own horse and was trained in cavalry drills so that even at a gallop he could keep up with the battery. The guns and vehicles themselves had teams of six horses instead of the usual four, and were expected to be able to accompany the movements of cavalry units. Horse gunners rapidly developed something of the *panache* and *élitism* of the cavalry, and would regard their companions of the foot artillery as boring stick-in-the-muds.

Almost as soon as the horse artillery had been formed it was recognised as an invaluable addition to the army, and its numbers were dramatically increased. It could accompany rapid strategic movements better than the foot guns, and could give cavalry formations an element of solidity which they had previously lacked. After the cavalry had forced enemy infantry to form square, for example, a horse battery firing at close range could wreak havoc in the closely packed ranks.

Horse artillery was also useful when it was co-operating with infantry, either skirmishers or the masses behind. It was more ready to go forward with an attack than the foot artillery, although as the wars went on the foot guns increasingly followed this example. It was pointed out that the apparent disadvantage which light guns might experience when advancing into the muzzles of heavier batteries could often be negated by the faster rate of fire of the former. On many a field the French horse artillery seemed to prove the truth of this, although it sometimes paid heavily.

Paradoxically horse artillery, which was so useful in the very forefront of a battle, was also ideal for use in the reserve role. When a general had developed his battle and was ready to commit his reserve at a weak spot identified in the enemy line, it was the horse artillery which he could most rapidly call up to prepare the way. For this reason the Guard Artillery contained a high proportion of horse batteries, and they often made considerable movements during the very climax of Napoleon's battles.

From what has so far been said it would seem that horse artillery was the ideal type of artillery, but there was another side to it as well. In the first place it shared with the Guard the disadvantage that it stripped the best men from the line and tended to become an *élite* speciality which was a law unto itself. The foot artillery naturally resented this, and there was less co-operation between the two than there should have been. Equally, horse artillery was extremely costly in the one asset which was in short supply to the French, horses. A horse battery had relatively poor firepower compared with a foot battery, yet presented a magnificent target to enemy gunners with its massed lines of horses and horse holders surrounding every gun. It was, paradoxically, slow to come into action, because although it could manoeuvre at perhaps twice the speed of a foot battery, it took longer for the gunners to dismount, park their horses, and run to their pieces. Indeed, such was the gymnastic rigour of the service that its gunners suffered heavily from that scourge of early nineteenth century medicine, an epidemic of hernias.

Particularly in night surprises the horse artillery was notoriously slow off the mark, as gunners and drivers would be milling round helplessly trying to decide which horse was theirs, tripping over tethering ropes, and tying each other in knots.

Weapon Capabilities

What could the French artillery weapons do? In the first place, of course, they made a loud noise, and this was of considerably more importance than is often realised. One of the most crucial calculations that a general had to make was whether his force was near enough to other forces for the sound of cannon to carry between them. Not only could concerted salvoes be used for signalling, but if a battle developed its noise was normally enough to bring all the detached corps rushing to support. This was almost a standard drill, and an isolated commander who failed to "march to the sound of the guns" could find himself in very hot water afterwards. The difficulty with this procedure, of course, was that it was entirely at the mercy of the wind, so no hard and fast distance could be quoted for how far the sound would carry.

The noise of cannon was also of great effect tactically. One of the main criticisms levelled against the four pounder gun was that its noise failed to terrify the enemy, whereas the eight pounder, and particularly the twelve pounder, made a very frightening noise. Their effect in battle was materially enhanced by this, so it should not be assumed that the importance of firepower in Napoleonic battles could be measured entirely in terms of the number of casualties produced. Gunfire also spread confusion and hesitation in the enemy ranks by its morale effect, and especially in the case of howitzer fire against cavalry this could be out of all proportion to the material damage caused.

Another effect produced by artillery fire was its smoke, although once again this was dependent upon the wind. After half an hour's firing on a still day this could add yet another element of confusion to the battlefield, obscuring the view of commanders and leaving troops vulnerable to sudden cavalry attacks out of the murk. At Friedland, too, during the early part of the battle the smoke was put to effective use by the French when they were able to conceal their true strength from the Russians, and by revealing the same troops successively at different positions behind the smoke screen they created the impression of far stronger forces than they in fact possessed.

The ammunition used by the guns was of several different types. Long guns would normally fire a solid shot against most targets, and on suitable ground these would ricochet at the end of their trajectory for a considerable extra distance. Thus enemy reserve formations could be engaged with the same shot as was initially aimed against the front line, and the beaten zone might extend for fully a mile from the muzzle of the gun. Towards the end of their flight these shots would be perfectly visible to the enemy against whom they were aimed, and there was many a hopeful footballer who lost a foot in a rash attempt to field them.

All artillery could fire grape shot and canister, which consisted respectively of about 40 large or 80 small balls in a tin box. When fired these would fan out from the muzzle in a destructive pattern which might reach out to about 600 metres, although experienced gunners would not hope for many hits beyond about 200 metres. This type of ammunition was for use against a broad frontage of enemy troops rather than echelons in depth, but in a moment of crisis at short range it could be every effective indeed. It could also be double loaded, or loaded in conjunction with a roundshot in order to increase the devastation.

Howitzers could fire rather larger doses of canister than guns because of their wider bore, but were usually used for indirect fire with shells. With a good system of observation they could fire from hidden positions behind the crest of hills, although with nothing approaching the degree of sophistication available to modern indirect fire weapons. The procedure was slow, for the shell had to be lit while it was in the gun barrel, and misfires were frequent. When it landed among the enemy there was often a long pause before the fuse burnt through to ignite the charge, and this might allow the nimble footed to take cover. Even when the shell did explode it fragmented into very few pieces, so that although any given splinter might carry for 20 to 40 metres, you had to be unlucky to get in the way of one. Napoleon demonstrated this to a unit of panicky conscripts at the battle of Arcis by riding his horse over a shell which was fizzing on the ground before it exploded. The horse was killed, but Napoleon wasn't. As an anti-personnel weapon shells appear to have done more damage by their initial flight than by the actual explosion of the charge, and would normally be used instead against buildings or entrenchments where direct fire was ineffective, or against ammunition supplies which they could destroy by secondary explosions. Their morale effect against cavalry has also been mentioned, so it would be wrong to suppose that howitzer sections could rarely be used.

As with so many aspects of Napoleonic warfare, the effect of artillery varied considerably with different conditions of weather and terrain. When the ground was swampy or soft after rain the ricochet effect was lost – not only from roundshot, but also from canister and shells. This could easily reduce the casualties inflicted by a half or even more, as was most notably the case at Waterloo. Conversely, ricochets would be much improved on stony ground where they would kick up extra pebbles into the enemy's face. Fire directed into villages was also particularly murderous, for splinters from the buildings would fly in all directions. Before the battle of Valmy, for example, the French gunners demolished the windmill which was in the middle of their line for fear of the danger which it represented in this respect.

Again, on a very hot day there was a further danger from the artillery, and this was fire. At Wagram the burning wadding which fell from the muzzles of the guns after each discharge started several huge fires which raged uncontrollably and burnt to death many of the wounded soldiers who were scattered over the field.

All these variables make it extremely hazardous to offer figures even for the range of guns. Theoretically, with a 45° elevation most field guns could fire up to four kilometres, but their carriages were only designed for elevation up to about 8° from horizontal. In battle, therefore, shots might carry 1½ to 2 kilometres, especially with a good ricochet, but at that sort of range the problem was visibility. For any degree of accuracy the effective ranges might be something like those cited by Guibert,[3] although in bad conditions these might easily be halved.

As for the accuracy to be expected, that is a very much more baffling problem. We hear, for example, that at Corunna some Spanish gunners tried to snipe at individuals with a 32 pounder gun, and the shot which wounded Marmont at Salamanca is legendary. Normally, however, firing at individuals was a waste of time, and even a screen of skirmishers

2 Guibert's Estimate of Battle Ranges (in metres)

Gun	Ball	Grape	Canister
12 pdr	900-1000	500-700	500
8 pdr	800-900	400-600	400
4 pdr	800-900	300-500	300

was considered a poor target. Artillery was for use against masses, but there again the accuracy is still hard to judge.

If we look at the results of peacetime tests (see diagram 4) we find a startling comparability between the accuracies claimed by the artillery of various nations. About one shot in three might be expected to hit a company sized target at long battle range, so assuming the enemy were drawn up three deep this would represent one casualty per shot fired. At canister range Lauerma cites the results of French tests which imply even more devastating casualties (see diagram 3). A battery of guns could apparently hit half a battalion with a single salvo. None of this can bear any relation to the tactical realities of battle, however, so for a sane estimate we must think again.

The other type of evidence we have available is in the accounts of the battles themselves. This shows, however, that the only generalisation we can make is that "everything depends upon the circumstances and the weather". One has no precise figures for such things as the number and type of rounds fired, the range, the ground, or the quality of gunners and target alike. All one can say is that performance apparently varied enormously, and occasionally even approached the sort of figures achieved in range tests under ideal conditions. At the battle of Sacile, for example, an Austrian gun carried off a file of three men in each of three successive shots, but in its subsequent fire hit nothing. At Konigsberg in 1807 13 men were hit by a single roundshot, and at Hanau in 1813, nine. Against this, on the other hand, we have the case of Wagram where regiments

3 Lauerma's Figures for French Experiments using Grape and Canister against a 5.80 by 35 metre target.

Gun	Number of Balls	Range in metres	Hits
12 pdr	41	700	10-11
	112	600	20-25
		400	40
8 pdr	41	600	10-11
	112	600	25
		500	40
4 pdr	41	600	8-9
	61	400	21

4 Accuracy of Field Guns in various tests against approximate company sized screens

Country of test	Source of information	Range	Per Cent Hits
Belgium	Fallot	900m	20
Austria	Lauerma (pp Liechtenstein)	1000m	40-70
Prussia	Lauerma	800m	35
Britain	Hughs (pp Muller)	950m	26-31

Average per cent of hits 34.

bombarded all day by the full weight of the French artillery lost only one eighth of their strength. Yet again at Austerlitz 40 guns unmasked suddenly against Lannes' troops are reputed to have caused 400 casualties in three minutes, yet at Smolensk four battalions of Hessians in square suffered only 119 casualties to 12 guns firing for three hours.

Perhaps one of the most important factors in all this is to be found in the rate of fire. Gunners were unanimous that at long range shots should be loaded and fired very carefully and deliberately. The fatigue of the gunners must be remembered in this context, for in a long battle they might be expected to manhandle their pieces for 12 hours at a stretch. Slow fire also improved accuracy, saved ammunition, and in particular it prevented excessive heating of the gun. One shot every three or four minutes would not be at all unacceptable, and in sieges heavier guns might be fired only once every half hour. In a close range crisis, on the other hand, quite the reverse was true. Canister rounds could be shovelled through the lighter pieces at the rate of four or five per minute, or two to three for howitzers or the heavier guns. It was this as much as anything which made close range fire so deadly, although such a "mad minute" might occur only once or twice in a battle.

Before we leave the performance of Napoleon's guns there is one negative aspect which is worth considering, and that is the question of spiking. There is a widely held misconception that it was easy to spike a gun in a crisis, whereas in fact it was a lengthy business and required considerable preparation. It is true that a temporary job could be done by hammering a nail or wooden plug into the firing vent, wedging a roundshot into the breech, smashing the guncarriage, removing the loading implements, or blowing up the ammunition; but all this still left a recognisable cannon in enemy hands which could be put back into service without too much trouble. To make a permanent job it was necessary literally to destroy the gun barrel itself. This could be done by burying the muzzle in earth and then firing the piece on a slow fuse, or by firing a heavy cannon at it from point blank range, as the French did in Almeida. Another method was to heat up the barrel while it was still resting on its carriage, and then drop a heavy weight onto it to break off the trunnions at either side. Finally the genuine process of spiking was not something which could be done by any amateur with a chisel, let alone a bayonet. It required a special hardened iron spike with a soft top. This was then hammered into the firing vent and the bottom bent round so that it could not be pulled or blown out. The top was then knocked off flush with the top of the gun.

Battery Organisation and Drills

The fundamental tactical unit of the artillery was the battery, which the French referred to as a "*Division*". By the late Napoleonic period this usually consisted of six guns of the same type plus two howitzers, although for the horse artillery there would only be four guns and two howitzers. Each piece would have its own team plus its reserve caissons, of which there would be three for twelve pounders and howitzers and two for other guns. These would carry a total of about 170 shots per gun, although there would be as much again in reserve in the army park. Guard Artillery batteries carried a double ration of ammunition, with about 350 rounds per gun packed in three caissons to each medium gun and five to each twelve pounder and howitzer. Napoleon reckoned his armies should carry enough ammunition for two good battles, and was particularly careful to maintain supplies.

In addition to the guns and caissons each battery would include one spare guncarriage and team, one mobile forge, one vehicle for tools and spare parts, and possibly also some caissons of infantry ammunition. Thus there might be a total of about 30 vehicles in the average battery, representing perhaps 140 horses. The manpower of the battery would consist of one company of gunners organised in four sections each of two guncrews, and one company of drivers from the Artillery Train. The total might be around 130 men. On the road this made a very long procession, perhaps 400 metres long with an allowance of 12 metres for each vehicle and one metre for the intervals between them. Guard and heavy artillery batteries would be even longer, and it was a recurring lament among Napoleonic generals that the number of vehicles in the army seemed to be continually escalating.

It is an amazing fact that until 1809 the French possessed no drill manual for manoeuvring their batteries, and even then there was only an unofficial one published by a group of generals who had been impressed by the number of guns used at Wagram.[4] It seems that drill manuals are the worst of all possible sources for information about what happens in battle, for they are almost always out of date. While the French in the Revolutionary Wars were using artillery in separate batteries they had no handbook which dealt with units higher than the individual gun. Yet as soon as they started to use multiple batteries they felt it was time to issue a drill only for the individual battery, still without official principles for the use of guns in mass. This was noticed at the time, but the repeated call for studies of higher tactics was satisfied only some years after Waterloo. This is not to say that French gunners were ignorant or inept at handling their pieces, but rather that they despised the pedantry of formal drills, and preferred to rely upon experience and a long practice.

On the march a battery would be in single or double column, depending on the width of the road, with the guns grouped together in front. In the case of the one or two light guns which might be attached to an infantry battalion the normal drill was to march in the interval between the first and second company. In the case of an army corps on the march some horse guns would be attached to the vanguard, and the mass of the artillery would be split up, with batteries near the head of each infantry division and a corps reserve somewhere in rear.

More than the other arms artillery required frequent rests of about ten minutes to close the intervals in the columns and to rest the horses. Even so it could sustain a speed of about three kilometres per hour on good roads, although it was naturally more affected by bad conditions than the infantry or cavalry, who could frequently march beside the roads. In the 1814 campaign, for example, the artillery was often floundering about in mud up to knee height, and would usually arrive at the night's camp long after the infantry. With the large numbers of vehicles involved, too, it required a lengthy process of marshalling to put a corps artillery park on the road at all, and it could easily take a whole hour to get 300 vehicles lined up.

Before starting a march an experienced battery commander would ensure that there was always an escort allocated to him of about two infantry companies. These had an obvious rôle in defending the guns from a surprise attack, but they were also extremely useful when there was no enemy about. They would reconnoitre ahead to find difficult defiles and fords (the Gribeauval equipment could wade in two and a half feet of water). They could widen the road if necessary, and help to push the vehicles on slopes or in mud. In the case of eight and twelve pounder batteries there was also a requirement for a certain number of unskilled labourers to help serve the guns in battle.

On arrival in the battle area batteries would be attached either to a reserve formation or the front line. In the former case they would be drawn up in a line of battery columns behind the reserve infantry, and would there take all the precautions necessary for coming into action. While in proximity to the enemy they would stand in readiness without unhitching their teams, even at night. By failing to observe this rule at Laon in 1814 some inexperienced naval gunners contributed significantly to the rout of Marmont's corps in a night surprise.

The batteries designated for posting to the front line would shake out from single file into column first of sections, then of half batteries, and finally into line as they left the road and approached the front. Before coming into range there would be an inspection of equipment and a general clearing for action. At this stage the guns might be put on the *prolonge* for ease of manoeuvre. Unless the battery was being deployed as part of a large mass of artillery, the battery commander would then make his reconnaissance. This was usually the most important part of the whole operation, for if a faulty battle position were chosen the results could be disastrous. With good siting, on the other hand, a single battery might be able to dominate two or three times its own numbers.

The first step in the reconnaissance was to discover the intentions of the local infantry commander; the general direction in which he wanted the guns to act, and whether the battle was to be offensive or defensive, mobile or static. It was always a cardinal point of artillery tactics that whatever else happened, the guns must act in the closest possible co-ordination with the other arms. Conversely, the artillery would always expect to be supported effectively in case it was itself charged.

Having seen the general situation the battery commander would then pick his position very carefully. This would be done with the aim both of obtaining the best field of fire and of protecting the battery itself. Plenty of open space in front of the guns was essential, with no dead ground and preferably good firm terrain towards the enemy so that ricochets could be bounced into his second line

5 Battery Formations

Single Column (Column of guns assuming six 6 pounders, and two 5 inch howitzers)

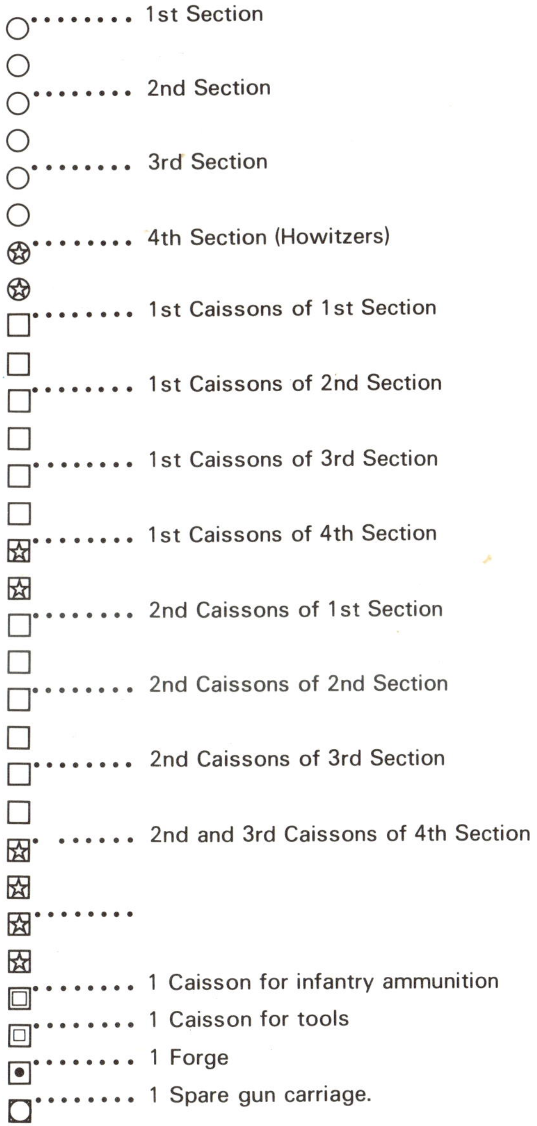

○ .... 1st Section
○
○ .... 2nd Section
○
○ .... 3rd Section
○
⊛ .... 4th Section (Howitzers)
⊛
□ .... 1st Caissons of 1st Section
□
□ .... 1st Caissons of 2nd Section
□
□ .... 1st Caissons of 3rd Section
□
☆ .... 1st Caissons of 4th Section
☆
□ .... 2nd Caissons of 1st Section
□
□ .... 2nd Caissons of 2nd Section
□
□ .... 2nd Caissons of 3rd Section
□
☆ .... 2nd and 3rd Caissons of 4th Section
☆
☆ ....
☆
▣ .... 1 Caisson for infantry ammunition
▣ .... 1 Caisson for tools
⊡ .... 1 Forge
○ .... 1 Spare gun carriage.

A Gunner 1804-1806 service dress.

B Drummer 1809-1812 campaign dress.

C Drummer 1804-1806.

D Gunner 1813-1814.

Foot Artillery of the Line

A Officer 1813-1814 campaign dress.

B Officer 1809-1812 service dress.

C Drum-Major 1st Regt. 1807-1812 campaign dress.

D Officer 1808-1814 Peninsula campaign dress.

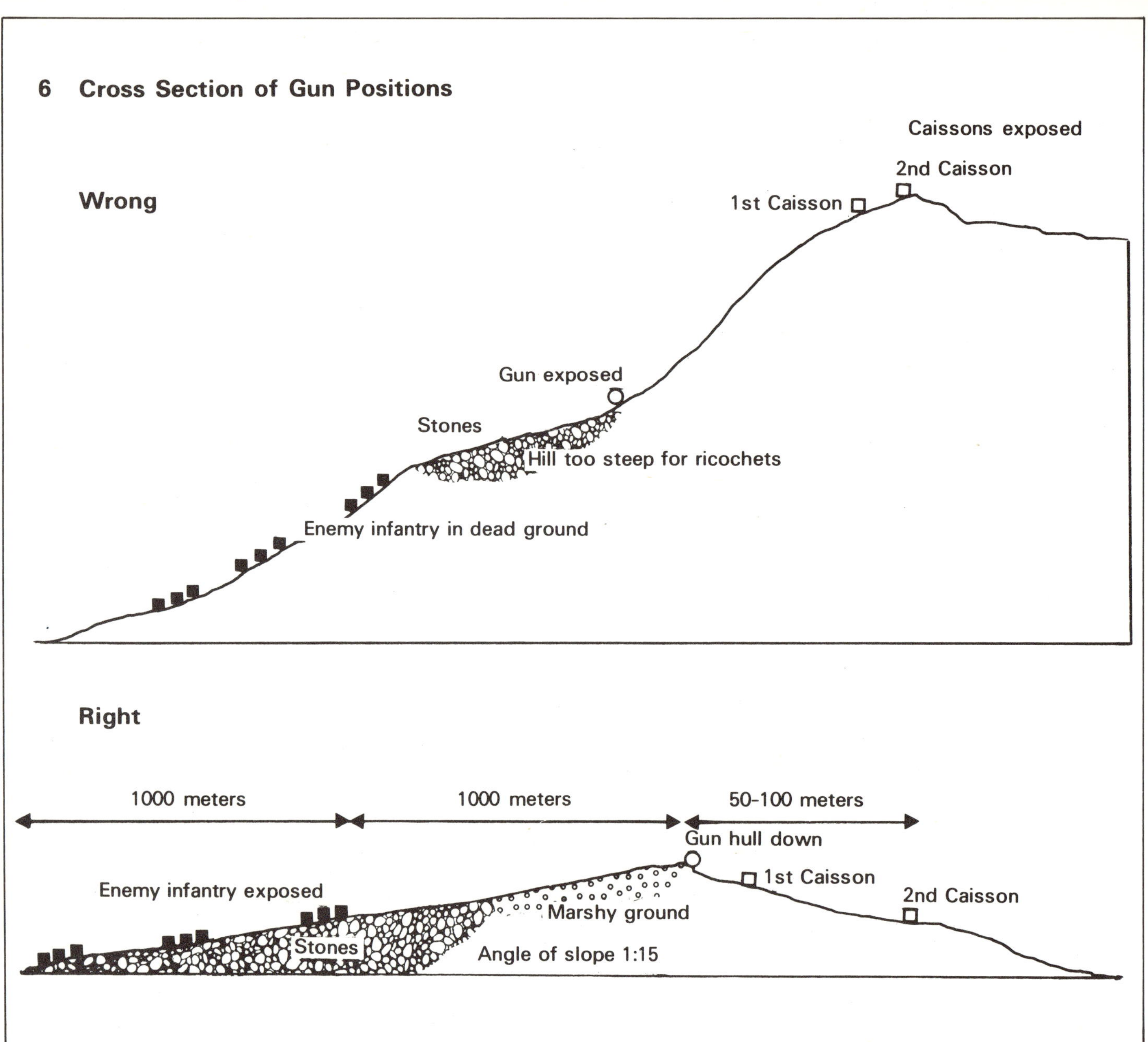
6 Cross Section of Gun Positions
Wrong
Caissons exposed
2nd Caisson
1st Caisson
Gun exposed
Stones
Hill too steep for ricochets
Enemy infantry in dead ground
Right
1000 meters
1000 meters
50-100 meters
Gun hull down
1st Caisson
2nd Caisson
Enemy infantry exposed
Marshy ground
Stones
Angle of slope 1:15

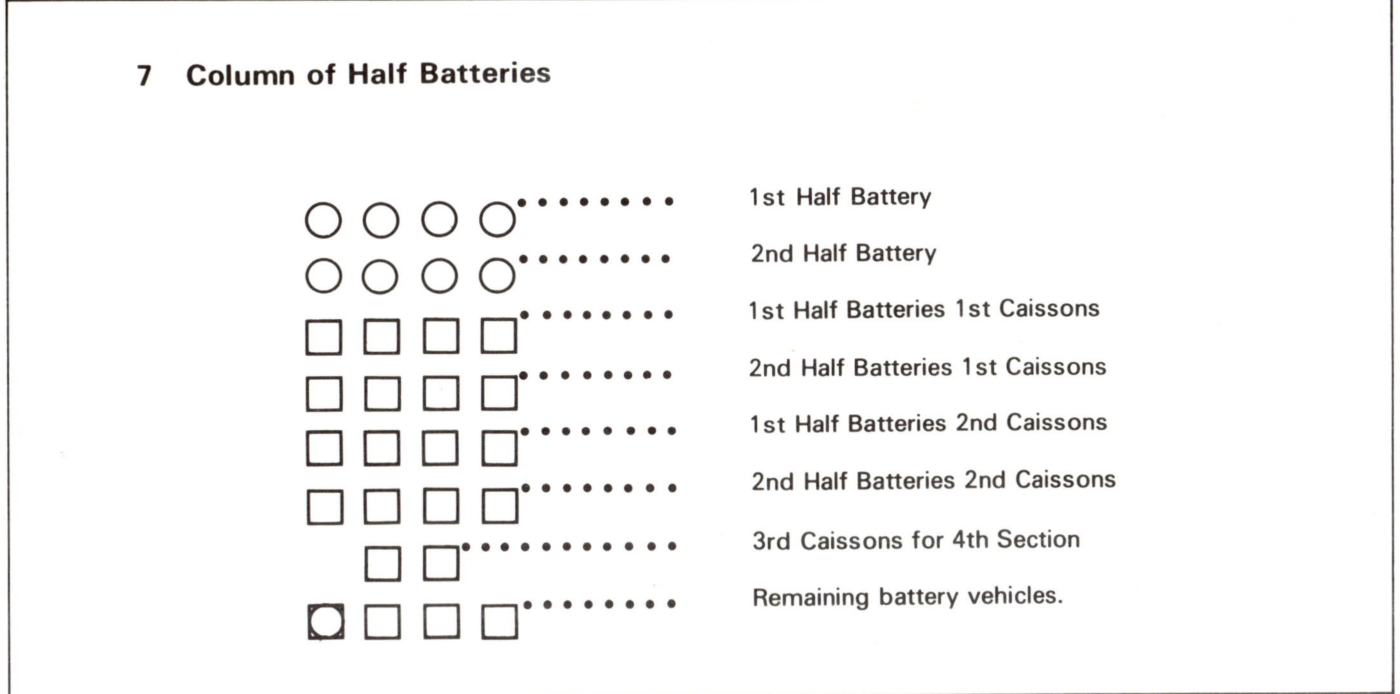

and reserves. A gentle slope away from the battery position was ideal, provided that it was no steeper than one in fifteen. Anything steeper than that interfered with ricochets and made it difficult to depress the guns far enough. In this context it is perhaps relevant to note that Wellington's favourite hilltop positions, as a Vimeiro, Bussaco, or St. Pierre, were so precipitous that they must have been an artillerist's nightmare. Just as the enemy came under long range fire he would disappear from sight beneath the convexity of the slope.

The protection of the battery was perhaps even more important than a good field of fire. Near rocks, trees, or buildings there was a great danger that the enemy's fire would kick up a hail of splinters, so swamps in front of gun positions were much sought after. If the guns could be placed "hull down" behind the crest of a rise it would also be a great help, although if this were not practicable low earth banks might be thrown up in front of the battery. This technique was learnt from the Russians, who were expert at all forms of fieldworks, but it is not clear how often the more mobile French artillery was able to make use of it in battle. Even better, of course, was the sudden unmasking of the battery from behind cover or a screen of friendly troops, as this would ensure both initial protection and then a surprise for the enemy. Especially in outpost work this could be an effective way of luring the enemy's supports forward into a fire trap, although on the grand scale it was also used at Lutzen and Hanau by the Guard Artillery reserve.

Protecting the guns themselves was headache enough for a battery commander, but their

Foot Artillery of the Guard

A Gunner 1810-1811 service dress.

B Sapper 1808 service dress.

C Gunner 1813-1814 campaign dress.

D Drummer 1810-1811 service dress.

Foot Artillery of the Guard

Officer 1810-1811. service dress.

Officer 1810-1811.

Gunner's greatcoat.

A Corporal 1810 winter society dress.

B Colonel 1810-1811 full dress.

C Drum-Major 1810-1811 full dress.

D Officer 1808-1810 campaign dress.

8 Battery in Firing Line

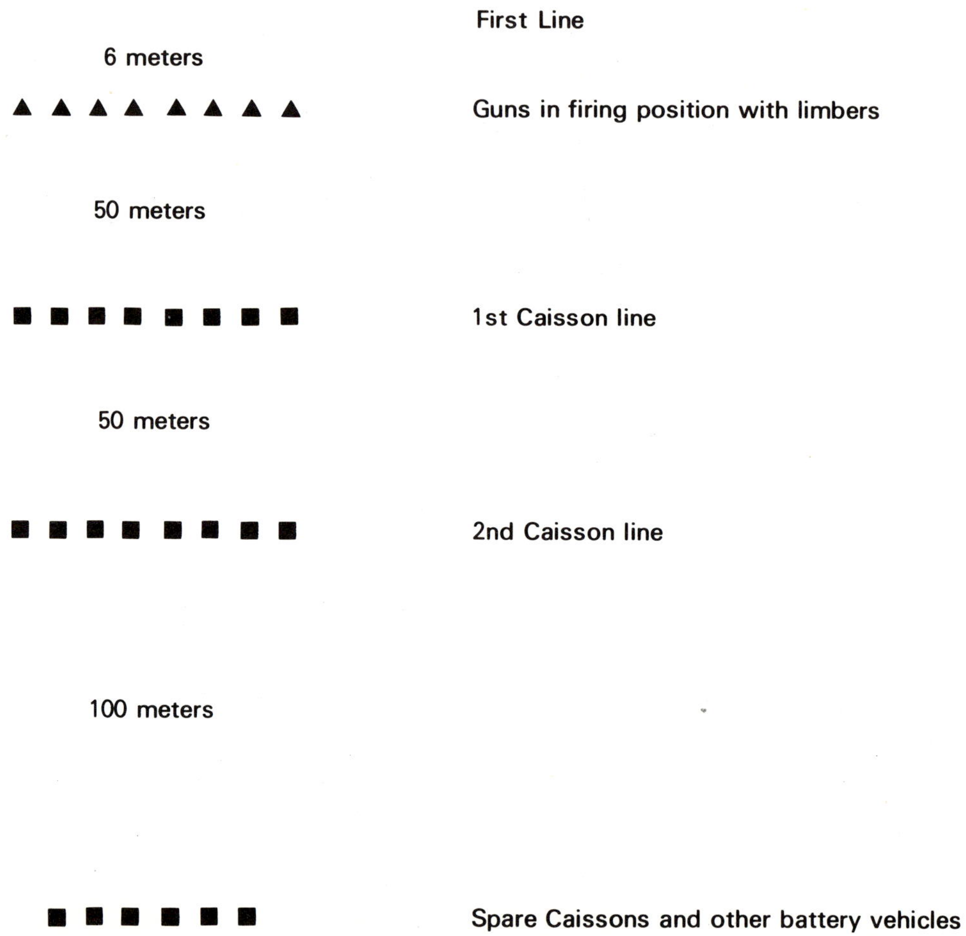

First Line

6 meters

▲ ▲ ▲ ▲ ▲ ▲ ▲ ▲ Guns in firing position with limbers

50 meters

■ ■ ■ ■ ■ ■ ■ ■ 1st Caisson line

50 meters

■ ■ ■ ■ ■ ■ ■ ■ 2nd Caisson line

100 meters

■ ■ ■ ■ ■ ■ Spare Caissons and other battery vehicles

supporting vehicles were more vulnerable. The ideal battery position would therefore have a concentration area for the caissons a short distance in rear which could be seen from the battery but not by the enemy. Whenever more ammunition was required caissons could then come forward one at a time to supply the whole battery rather than exclusively the guns to which they were theoretically attached. Yet further behind the first caisson line the empty caissons and battery vehicles which were not immediately needed would be parked under command of the second captain of the battery, who would keep a distant watch on the battle from a safe position and move in conformity with the battery's movements.

Other precautions which the battery commander might have to take would include a special watch against the danger of fire if his approach march passed any buildings which were burning, or even which had a fire in the hearth. A single spark might be enough to set off a caisson. He would have to ensure, too, that his battery was not posted directly in front of other troops, for the shots he might attract would continue on their way to hit these. Finally, the battery position itself would have to have plenty of room for manoeuvre, a clear exit, and preferably an alternative fall-back position if things got too hot.

Having decided on his position the commander would post markers to decide the specific site of each gun and to estimate ranges. The guns would then be summoned and would advance to their posts at a walk (86 metres per minute), a trot (189 metres per minute), or a gallop (200 metres per minute). The more dangerous the position the faster they would come, in order to return fire as quickly as possible. In practice they would be placed about 12-20 metres apart, although in theory it should only have been six metres. The greater the spacing, the smaller would be the target for the enemy, and the easier it would be to manoeuvre. Only in a great massed battery as at Wagram or Leipzig would the guns be placed closer together, to fit them into the space available. Batteries would also normally avoid deploying along a straight line in order to minimise the danger of enfilading fire against them. Instead they would prefer a ragged, broken line which would have given a drill instructor apoplexy. The same drill instructor would no doubt have suffered a relapse when he realised that in their deployments the artillery were quite unworried about keeping

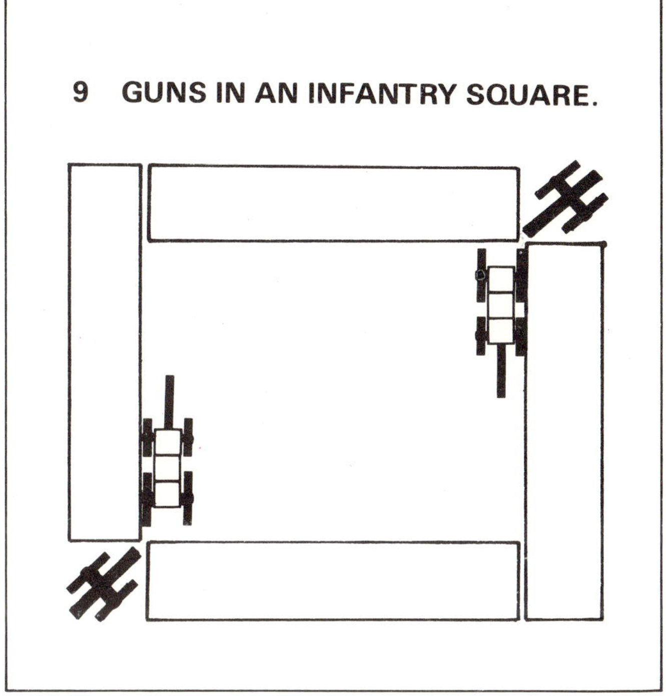

sections in numerical order from the right, and dispensed with all the paraphernalia of *"inversions"* which so plagued the other arms.

When the guns arrived at their posts there would often be an embarrassing pause; in the case of horse artillery rather a long time was needed for the gunners to dismount and take up their stations, while for eight and twelve pounder batteries it took perhaps three quarters of a minute to move the gun barrels from the travelling to the firing trunnion positions. Meanwhile the vehicles would be taking their stations - always facing to the rear - and the *"coffrets"* would be opened for the ready ammunition. When all was ready the gun crews would go to their posts (see diagram 10), and fire would be opened. This would rarely be in salvoes, as that would leave too long a pause between discharges,

Horse Artillery of the Line

A Gunner 1807-1808 service dress.

B Trumpeter 2nd Rgt. 1809 parade dress.

C Gunner 1812-1815 campaign dress.

D Trumpeter 1812-1814 Peninsula Dress

A Officer 4th Rgt. 1804-07 Service dress.

B Officer 1810-12 Winter campaign dress.

C Trumpet-Major 1809-10 Campaign dress.

D Officer 1811 Service dress.

but normally by the two guns in each section alternating their fire when they saw their fellow had reloaded. In a crisis, of course, fire would be fast and furious, and by the use of *"bricole"* and *"prolonge"* the guns could be moved quickly into new positions if need arose.

The tactical uses to which a single battery could be put were many and various. In defence it would be a powerful deterrent to enemy attacks, and the gunners would always try to fire into opposing infantry and cavalry masses rather than artillery if it were at all possible. In this way not only would enemy attacks be broken up and disordered by the time they arrived within decisive range, but the friendly infantry would also be reassured to hear the noise of their supporting bombardment. Particularly with raw troops most generals recognised that battalion guns at least achieved the latter effect even if they were not particularly helpful in a material sense.

In the defence of villages the guns would be placed among the houses themselves only if there were some very strong cover available: a stout chateau or a solid cemetery wall. Normally artillery was posted on the flanks in order to continue an enfilade fire right up to the last minute. In open country, equally, the guns would always try to reach the flanks of an attack, and might even advance to such a position while the enemy was actually in motion. Cross fires and enfilades were for ever foremost in gunners' minds, and in the Napoleonic Wars the French became expert at achieving them.

Against cavalry the artillery was vulnerable, and it would seek refuge in its own infantry squares, usually firing from the corners for the widest arc. It is a myth that guns often repulsed cavalry attacks with their fire alone, although technically it was quite possible. In practice the element of fear was usually stronger, and it was only a rash battery commander who would rely upon his men to continue firing canister until the very last minute. There are certainly many examples of this happening, and even of the gunners beating off attacks hand to hand. Twice in 1800 in Germany some horse gunners went to the lengths of mounting their own horses and countercharging cavalry on its own terms. Such incidents, however, were always very much the exception, and it was more usual for an overrun battery to fall into enemy hands.

When artillery was used to support an attack it would once again attempt to fire into the enemy's flank, both to obtain an enfilade and to avoid being masked by the assault troops. The closer the artillery could come, the better; and the French became very bold in this role. It required quick reactions, however, to guard against sudden counterattacks so near to the enemy; and sharpshooters were a continual menace. Particularly in the Peninsular War the French lost many guns and gunners as a result of

Position of Gunners

A Aimers and gun commanders
B Loaders and ammo feeders
C Firers
D Avant-train and ammo handlers
● Extra gun handlers from infantry.

10 Firing 1

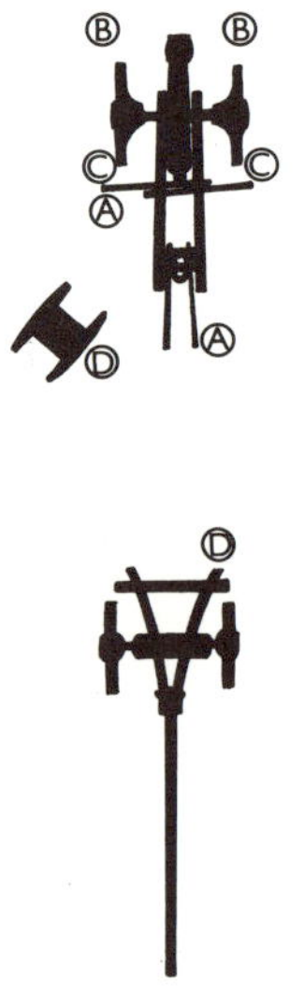

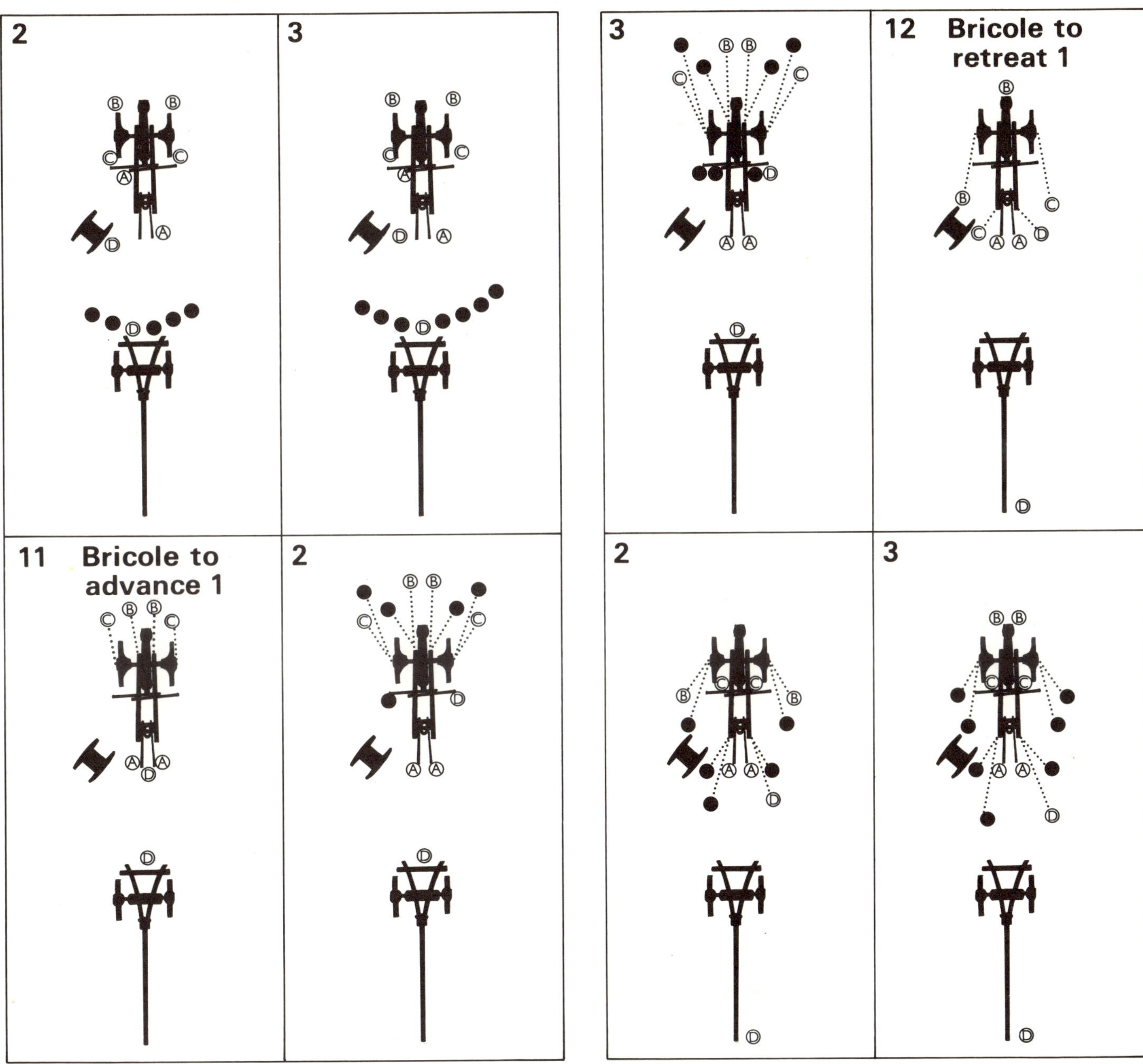
2
B
B
C
C
A
D
A
D
3
B
B
C
C
A
D
A
D
11 Bricole to advance 1
C
B
B
C
A
D
A
D
2
B
B
C
C
D
A
A
D
3
B
B
C
C
D
A
A
D
12 Bricole to retreat 1
B
B
C
C
A
A
D
D
2
C
C
B
B
A
A
D
D
3
B
B
C
C
A
A
D
D

Horse Artillery of the Guard

Corporal winter campaign dress.

Trumpeter 1802-05

Sergeant 1812-14 campaign dress.

A Trumpeter 1806-1814 Parade dress.

B Gunner 1806-07 Service dress.

C Trumpeter 1806-14 'Petite tenue'.

D Gunner 1812-14 campaign dress.

A Officer, 1809
Service dress.

B Officer
Summer service dress.

C Officer
Parade dress.

D Sergeant
campaign dress.

coming too close to unsubdued enemy light infantry. To some extent the danger could be minimised by leapfrogging forward by alternate sections or by approaching behind a screen of troops and then deploying suddenly to one side. The latter tactic also increased the surprise value of the fire and could often rout an inexperienced enemy on its own.

A more certain method, however, was the one which we shall now consider, and the one which was increasingly relied upon in the later Napoleonic Wars. This was the use of artillery in large masses.

Massed Artillery

We have seen that the individual battery was in itself a powerful and tactically flexible weapon. In support of an infantry brigade it could materially assist the local commander's battle. When the divisional system was introduced, however, the bulk of the artillery was eventually transferred to this higher level where it could be concentrated under the divisional commander in formations of two or three batteries. The divisional allocation of artillery later settled down at one medium and one horse battery.

In the battles of the Revolutionary Wars there were thus often 18 to 24 guns massed at decisive points, and it was soon noticed that these exerted a greater influence than the sum of their component parts. At Castiglione, for example, Marmont forced the key to the Austrian line with 19 guns, and at Marengo he stopped the decisive final attack with 18. It was becoming increasingly clear that concentrations of this type were the best way in which artillery could be used.

At Austerlitz and Jena artillery concentrations continued to be about the size of a single divisional park. One factor which was being added at this time, however, was the growth of the Guard Artillery as an effective army reserve. At Austerlitz 18 heavy reserve guns (not from the Guard) were gathered on the dominating Santon knoll in the long-accepted role of "guns of position". The static use of an army reserve in this way was well known in all armies and caused no surprise. What caused rather more interest, however, was what happened when the corps of Lannes and Soult became separated and a gap appeared in the middle of the French line. The Guard Artillery raced up to fill it with 24 light pieces, and successfully kept the opposing masses at bay.

At Jena the battle was an almost unprepared encounter and the artillery was not concentrated until once again a gap appeared in the French line. This time it was caused by the separation of the corps of Lannes and Augereau. Lannes sent 25 guns to cover the gap, and a dangerous moment was averted. It should be noticed, however, that although Napoleon had gone to great lengths to bring the Guard Artillery onto the battlefield, and kept it ready to move at a moment's notice, on this occasion it was not used.

Apart from the appearance of the Guard Artillery at this time there was also the effective organisation of the army into corps, each composed of several divisions. Once again the gunners thought they saw the chance to make even heavier concentrations than before, and dreamed of corps artillery commanders being able to unite the guns from all their divisions, together with a corps artillery reserve. This would make a mass of between 36 and 50 guns; obviously an improvement on the resources of a single division. Especially in the 1807 campaign against the Russians it was noticed that the enemy was also in the habit of massing his guns, so an increment of this type would have been doubly welcome. Unfortunately there was an almost insuperable obstacle in the way of this development, in the form of the jealousy with which divisional

A Soldier 1813-1814 campaign dress.

B Trumpeter 1809 campaign dress.

C Sergeant 1805 parade dress.

D Soldier 1813 parade dress.

Artillery Train of the Line

A Trumpet-Major 1813-14 campaign dress.

B Sergeant-Major 1813-15 Service dress.

C Lieutenant 1809 Full dress.

D Lieutenant 1813-15 Service dress.

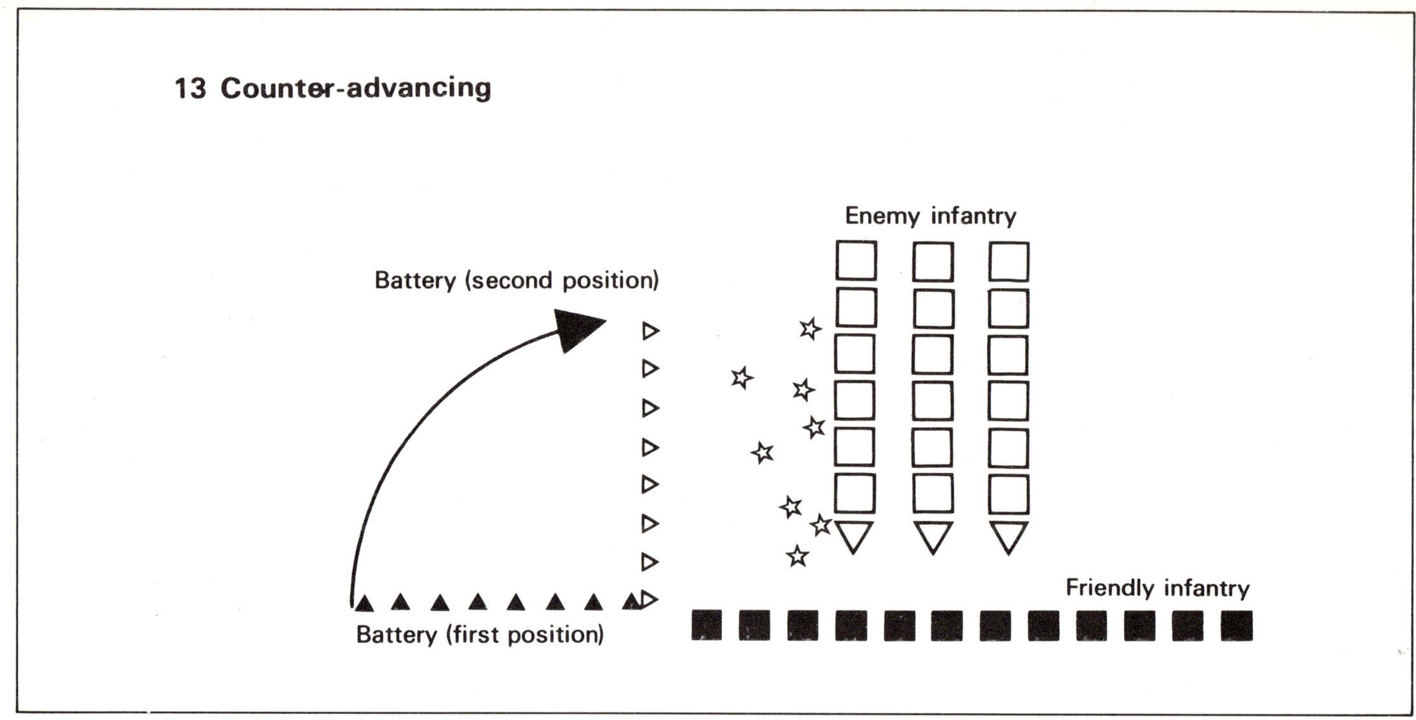

commanders would guard their guns against poachers from higher up. Fancy new ideas like corps artillery tactics were all very well, these men would reason, but without its artillery a division was locally vulnerable. It was only in exceptional cases, therefore, that the average line army corps would concentrate all its guns in one place. The corps reserve certainly added to the flexibility of the system, but in most of the remaining battles of the Empire the normal grouping of batteries was inside the individual division. At Orthez in 1814, for example, Soult defended each of three spurs with a divisional artillery concentration.

If the artillery concentrations of the line corps had reached their ceiling, those of the Guard Artillery continued to grow, and were free from the organisational limitations of the line. From the start the Guard Artillery had been intended to intervene *en masse* after the battle had developed, and hence it was not used in the type of desultory outpost work on which the line was often dispersed. At Eylau, therefore, the Guard was able to assemble 40 pieces in one place while St. Hilaire's division could raise only 18, and Sénarmont had to be content with 19.

This was to be the pattern for the later battles of the Empire, with the Guard providing the nucleus for ever bigger masses of guns. Especially after the great concentration of 100 guns at Wagram Napoleon became convinced that a really decisive result could be obtained only with 36 guns or more. Against this weight of fire, he said, "Nothing will resist, whereas the same number of cannons spread out along the line would not give the same results".

It was about this time that artillery tactics came of

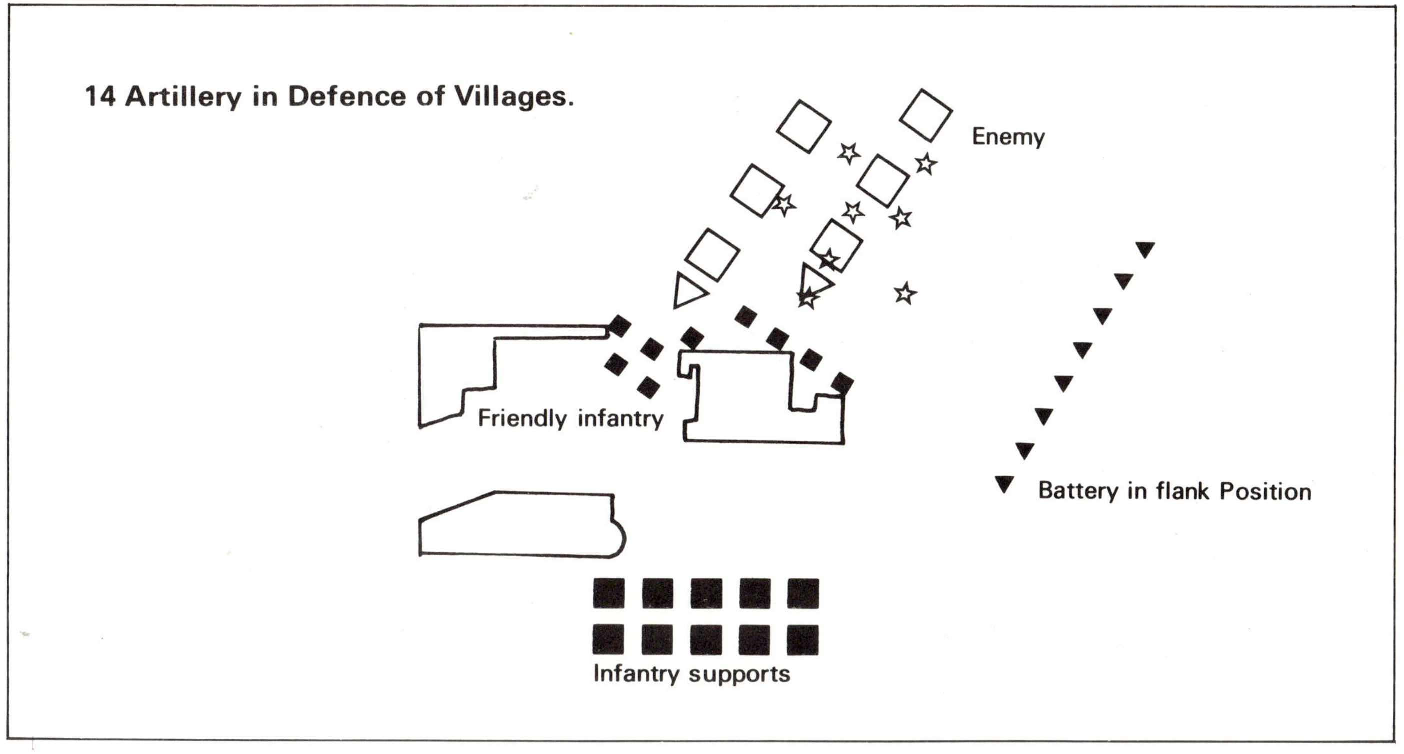

age. The artillery was no longer merely an auxiliary which assisted each division to achieve a limited result. Instead, the Guard Artillery mass could claim to be one of the great deciders of battles. "The object of artillery is not to kill men or dismount pieces in isolation but to make holes in enemy front, stop his attacks, and support those launched against him". To achieve this the secret was concentration of effort, and the ideal distribution of an artillery mass was in a semicircle around the point to be battered (see diagram 15). With this formation the guns would cross their fire and give the maximum chance of an enfilade. Each battery would also be well separated from its neighbours, and would thus offer a relatively small target to the enemy.

For a virtuoso demonstration of what this could mean in battle we must look at the classic performance of Sénarmont at Friedland in 1807. This was before the Guard Artillery had achieved the tactical preponderance it was later to enjoy, and was an exceptional case in that the artillery commander of a line corps was for once allowed to bring together all the guns under his command, thus stripping each division of its local firepower. Friedland was also exceptional for the mobility and opportunism with which the guns were handled. As we shall see later, these qualities often suffered when a more deliberate attempt was made to use guns in mass.

Like so many of Napoleon's battles Friedland had started as an unplanned encounter, and for a long time the French were fighting a desperate holding

Artillery Train of the Guard

A

B

C

D

Cpl.
1805-1806
parade dress.

Soldier
early 1812
campaign dress.

Soldier
late 1812
campaign dress.

A Soldier 1805-1806
campaign dress.

B Trumpeter 1815
campaign dress.

C Trumpeter 1809-1811
parade dress.

D Corporal 1813-1814
campaign dress.

Artillery Train of the Guard

A Officer 1805 parade dress.

B N.C.O 1803-1809 campaign dress.

C Trumpet-Major 2nd Rgt. 1813-1814.

D Officer 1809-1814 campaign dress.

15 Ideal Tactics for Massed Batteries

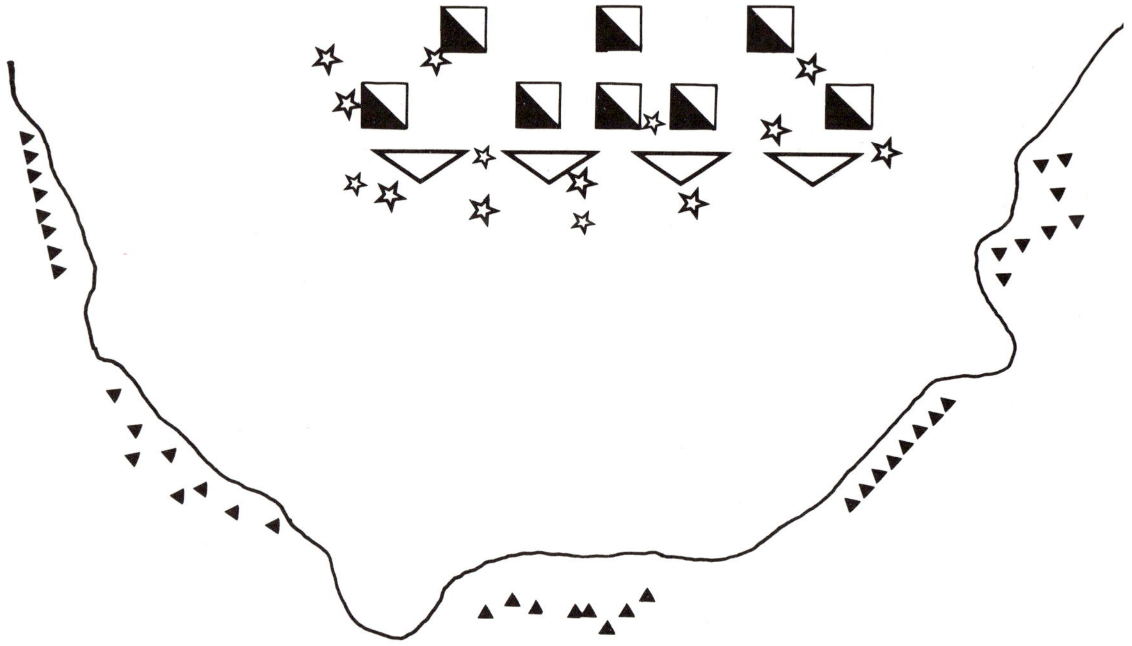

Enemy Infantry and Cavalry under crossed fires. Batteries widely spaced and in good covered positions.

action against superior numbers of Russians. Eventually, however, the French reserve masses came up, and were able to start an offensive to trap the enemy against the river Alle to his rear. On the right it was Ney's corps which led the attack, but he was soon halted by a cavalry counter attack and by fire from the far side of the river. As he fell back in confusion he dropped the battle into the lap of Dupont's division, which was leading Victor's corps. At this point there was only a single battery with Dupont, and although the line was held the position did not look particularly promising. It was then that Sénarmont, who was Victor's corps artillery commander, arrived on the scene. Initially he had merely been supervising the removal of some wounded horses, but he quickly saw that there was an opportunity for an artillery concentration. Rushing to Victor he obtained permission to mass the guns from all three divisions of the corps, a total of 38 pieces. These included four twelve pounders, four four pounders, eight howitzers, and 22 six pounders.

Sénarmont split his guns into three provisional batteries; a heavy reserve and two main units each of ten six pounders, two four pounders, and three howitzers. The two big batteries were placed on hillocks some distance apart to cross their fire, while the reserve was kept in a covered position behind the left hand battery. Fire was opened at 400 metres from the enemy, but after five or six shots from each gun the batteries were advanced alternately to about 200 metres. They were supported by one infantry battalion and four dragoon regiments, while the remainder of the corps sheltered behind a fold in the ground to the rear. So precarious did this advance appear, indeed, that Napoleon is reputed to have thought Sénarmont was deserting. It was a novel demonstration of the fact that artillery could make a charge on its own in the same way as the other arms, and Sénarmont was deaf to all attempts to call him back.

When they were 200 metres from the enemy line the French fired about 20 times, still with roundshot. By this time it was half an hour after they had entered the action, and they were beginning to dominate the situation. They were lucky, however, that the enemy guns across the river were unable to inflict many casualties upon them because the Russian field of fire was obstructed by both the proximity of friendly troops and the dense lingering smoke.

Sénarmont's next step was to *prolonge* both batteries forward until they joined together at about 60 metres from the enemy. A rapid fire with canister was then sufficient to break the infantry in front of them, and hence to silence the supporting artillery, for the Russian gunners did not stand their ground once their infantry had retired. The Russians then attempted a cavalry charge, but this was obstructed by fugitives and deterred by two general discharges by the French. After this the way was clear for a French infantry attack right into the town of Friedland itself. Sénarmont accompanied this for most of the way, and also poured fire into the flank of units which were retreating across his front to the safety of the Alle bridge.

Three hours after Sénarmont had come into action the battle had been won. In that time his guns had fired an average of 72 rounds each, plus twelve rounds of canister. It may seem surprising that although the entire action had taken place at such short range canister was not used more liberally, but the French gunners were probably reluctant to use it except at really point blank range. Against tightly packed masses, and particularly against the fugitives later in the action, solid shot was probably the most effective ammunition.

Sénarmont lost 66 casualties killed and wounded in this battle, plus 53 horses. If there had been less smoke on the battlefield, or if the Russians had deployed an effective light infantry screen the losses

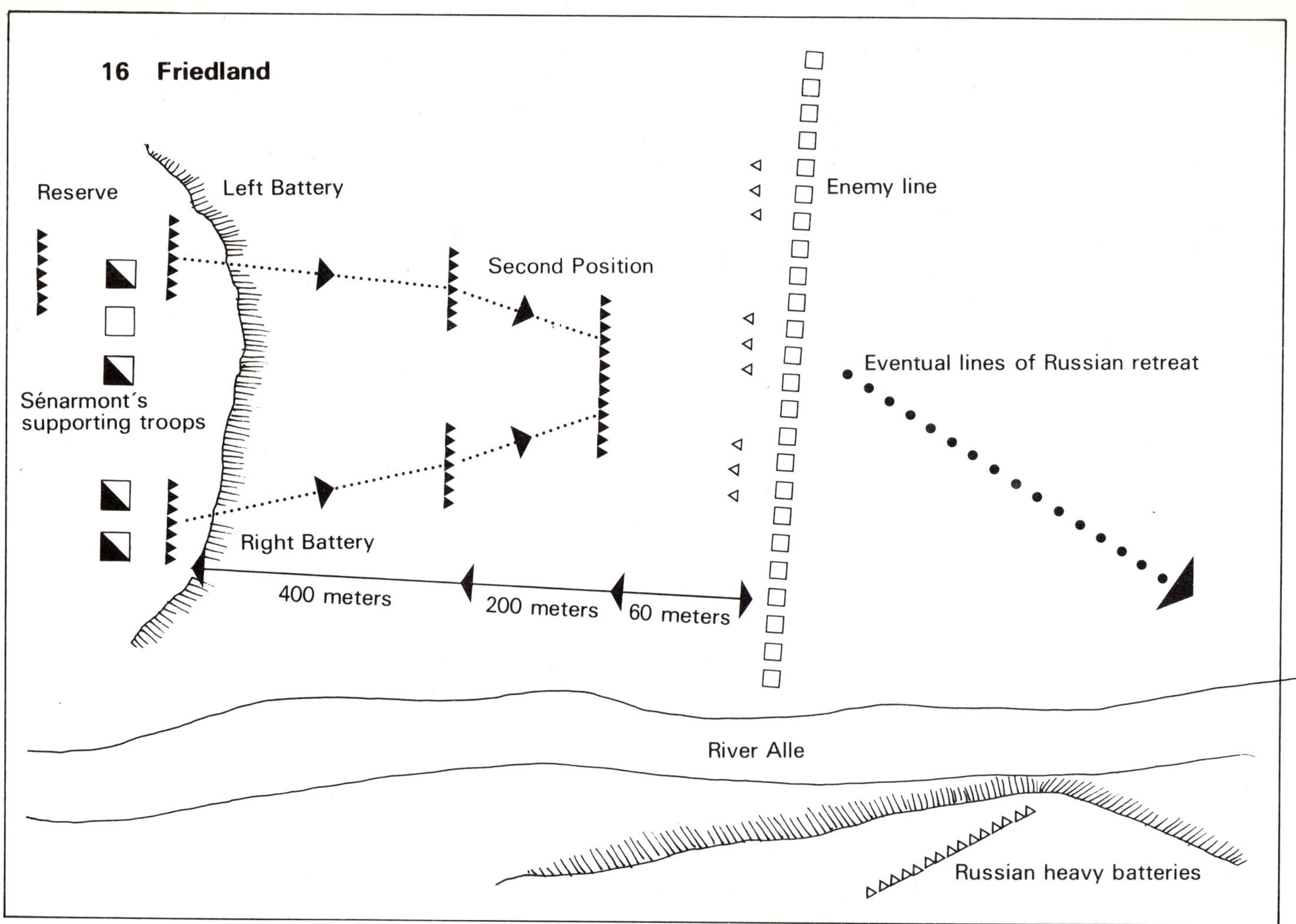

would undoubtedly have been far heavier, although it was perhaps the strong French cavalry supports which prevented the latter from happening. On the other side Sénarmont claimed there were 4,000 enemy left on this part of the field, although a proportion of these would have fallen in the earlier fighting and the subsequent infantry attack. Nevertheless the artillery charge was a startling feat of arms, and served to prove the old horse gunner Foy's dictum that "the best tactic is to get up close and shoot fast".

The Guard Takes Over

Friedland was an exceptional case, and although it was won by the concentrated artillery of an entire corps, the number of pieces was still not very much larger than had been used in previous battles. After this time Napoleon was starting to think in terms of masses of 80 or 100 pieces, and to find these he turned to the Guard. It became a standard tactic to assemble a force of this strength which could intervene at the climax of a battle in the direction thought to offer the greatest chance of decisive results.

The first spectacular example of this was at Wagram, where towards the middle of the day the Guard Artillery was brought up to support the hard pressed army of Italy. Together the two units could muster 102 pieces, and this artillery, deployed along a mile of front, succeeded in halting Kollowrath's advance against the French left centre. Napoleon then turned these guns to an offensive purpose, to clear a way forward to the villages of Aderklaa and Sussenbrunn. With Drouot's heavy batteries in the centre and d'Aboville's horse guns on the flank, the ponderous French mass engaged the enemy in a duel, creeping nearer to his line all the time. Unfortunately this was to be no repetition of Friedland, for the enemy was better prepared. He brought down effective fire on the French guns from both artillery and skirmishers. As the Guard horse guns rode up they lost 15 pieces even before they could get into action. It was true that within half an hour the Austrians had been silenced and much of their infantry had retired from its advanced positions; but this did not amount to a victory. There was both a desolation in the French gun line which had not been seen at Friedland, and a failure to make a rapid exploitation of the breach with infantry. When Macdonald's massive column finally got under way the effect of the bombardment had passed and his troops were stopped by a strongly posted reserve line.

Critics of the massed battery of Wagram have blamed its failure on several factors. There was an insufficient softening up of the enemy before the guns were committed at close range. There was poor co-ordination between Drouot and d'Aboville, who after all had no previous experience with masses of this size. Perhaps most important of all, the infantry exploitation was botched. Time and again in the battles of the later Empire it was to be this final weakness which marred the intervention of the Guard Artillery. With masses of this size the gunners could no longer view their operations from the level they were used to, but were forced to fit into an altogether higher framework of control and command.

It was one thing to co-ordinate a battery's evolutions with those of a regiment of infantry, but with 100 guns you needed at least an army corps on call, and usually more.

The use of this number of guns at a single point represented a deterioration from the delicate and flexible tactics of earlier battles. One could no longer keep guns widely dispersed to present a small target, for the restricted space would force them

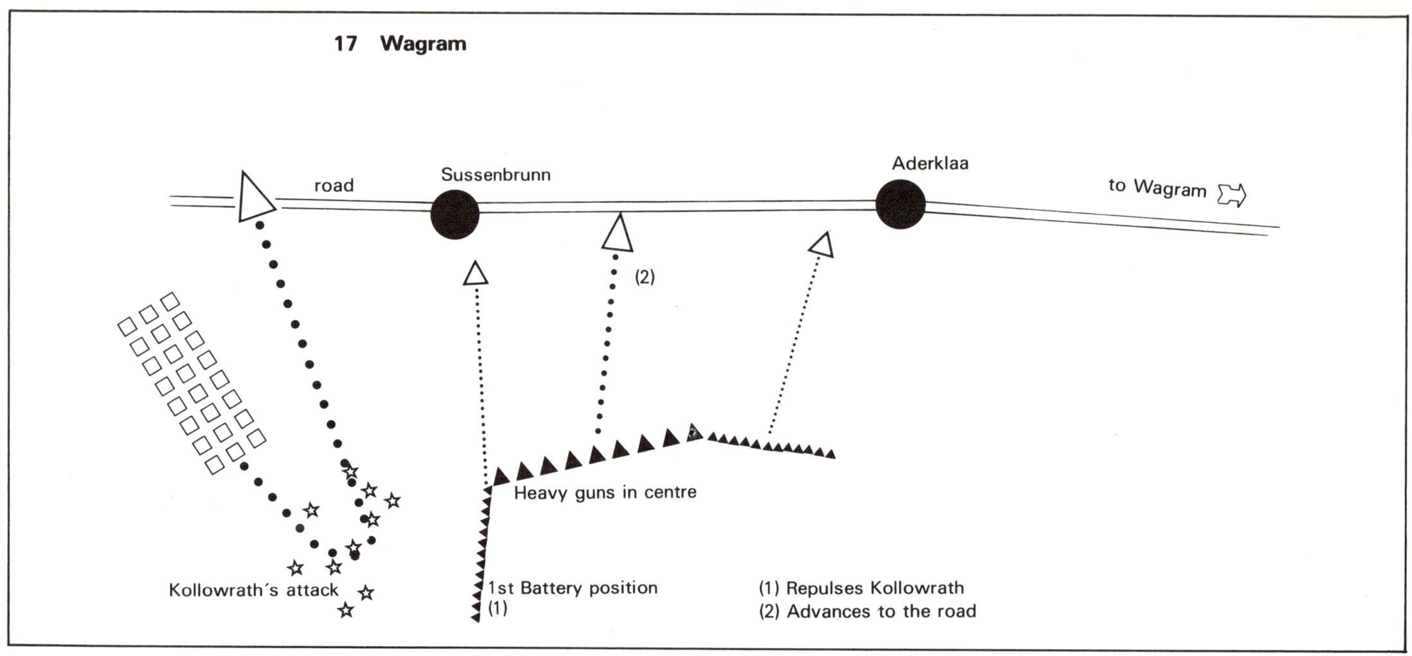

together. Fire could not always be crossed at the centre of a semicircle, and at Wagram the artillery was in fact in a convex line. On the other hand if the enemy were also using large masses of guns there was very little alternative but to answer him in kind. In 1812 at Borodino both sides deployed more artillery on a narrower front than had been seen even at Wagram, yet succeeded only in cancelling each other out. But at Borodino, of course, the Guard infantry was forbidden to exploit the break-in created by its artillery.

In 1813 the losses suffered in Russia had so weakened the army that Napoleon was forced to rely even more heavily upon his Guard. Despite a shortage both of experienced gunners and of horses, therefore, the Guard Artillery was increased, with particular emphasis on twelve pounders and the new heavy howitzers. Wherever Napoleon went his Guard Artillery went too, and it was surely no coincidence that the enemy started to plan battles only at points where Napoleon was not thought to be present. "It is the Artillery of my Guard", said the Emperor, "which decides most of my battles, because I am able to bring it into action where and when I wish".

A pertinent example of this was Lutzen, where 60 Guard guns appeared from behind a masking ridge; stopped the victorious enemy in his tracks; and prepared the way for an infantry counter attack. In this case, however, a new weakness in the use of mass batteries became apparent, since the infantry outran the beaten zone cleared by the guns. The enemy was able to solidify his defences beyond the range of the bombardment, and the French had to be content with only a limited advance.

At Bautzen Marmont's central attack was supported by 76 guns, but here again not as much was achieved by massed artillery as had been hoped. The

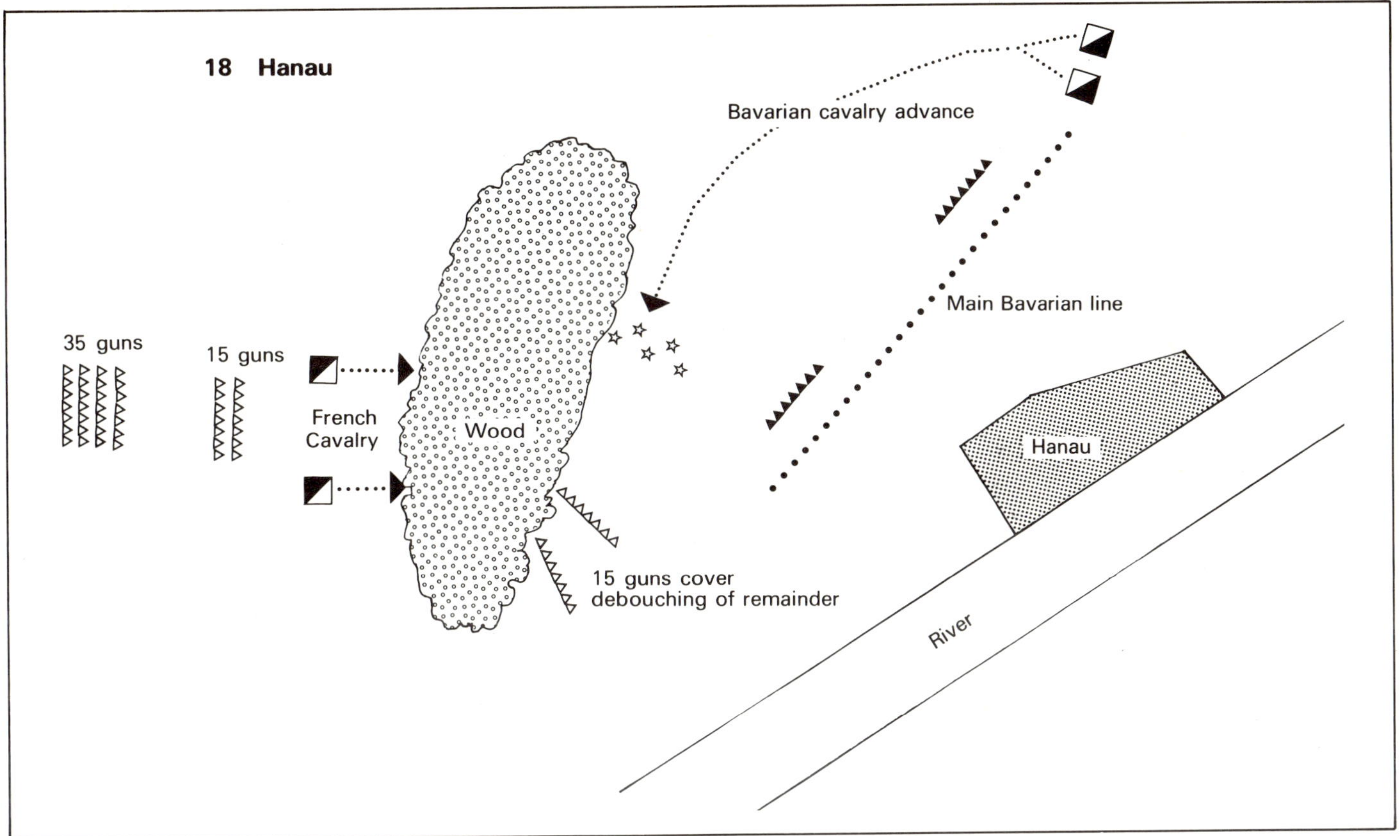

terrain was both too hilly and too soft for the fire to be very effective, and the Russians were able to preserve their batteries in fieldworks outside the effective range of the French guns. The French attack therefore fell foul of these guns, and bogged down into confused fighting in close country.

At Dresden the Guard Artillery was committed on the second day around the Grosse Garten in an attack on the enemy line. The guns came on in grand style by successive batteries, and cleared the enemy from a wide zone to their front. At this point, however, they were inexplicably ordered to retire, and in his memoirs General Griois records his fury at the waste of a good chance to break out. His rage reached a peak at Leipzig, where at Wachau on 16th October he suffered a similar experience. A Guard mass of 80 pieces, including 32 of heavy calibre, again blew a convincing hole in the enemy line, but the supports were not well organised, so there was again a failure to exploit the advantage.

A more successful use of the Guard Artillery came at Hanau, where the French army in retreat from Leipzig fought its way through a road block of Bavarians. At first the infantry was unable to make progress in some woods, but when Drouot arrived with the Guard Artillery his preliminary reconnaissance showed him a profitable line of attack. First he cleared the woods with two battalions of Guard infantry in skirmish order, then he defiled through them with 15 guns and deployed on the far edge of the woods, opposite the enemy's main line, but in such a position as to take his artillery in flank. Behind this screen the remainder of the Guard Artillery deployed, making a total of 50 pieces. When these guns had begun to throw the enemy into confusion the Guard Cavalry completed the rout. The Bavarian cavalry then counter attacked, but Drouot held his own cavalry in front of the guns until the very last minute. When the enemy had arrived at close range the French cavalry then wheeled away to reveal the massed battery, which proceeded to make very short work of the Bavarians.

In 1814 the Guard Artillery was able to make a decisive intervention at the crises of the battles of Craonne and Montereau, in both cases on extremely narrow frontages. Such was the speed of manoeuvre in this campaign, however, that the mass of guns came up too late for most of the other battles, or were obstructed on the battlefield by the thick mud. At Bar sur Aube, for example, the guns had to use double teams to move at all.

The final battle in which massed artillery was used was of course Waterloo, but after its spectacular success at Ligny the artillery was here less than effective. The ground was too soft for ricochets, and too treacherous for easy manoeuvre. Wellington kept his line in covered positions, while Ney's exploitation phase was mishandled. With large masses of all arms the essential co-ordination was always difficult to achieve, and it was perhaps Napoleon's greatest failing as a general that the problem had still not been solved by 1815.

The story of artillery masses cannot be rigidly confined to Napoleon's battles, however, for the technique was also applied in the Peninsula. Not, it must be said, with any great effect against the British, but certainly with a devastating effect against Spanish armies. Whereas Wellington would accept set-piece attacks only on ground which was eminently unsuitable for artillery, our old friend Sénarmont was twice able to effect a central breakthrough against the Spanish, at Oçana and Medellin. The French also used a mass of 60 guns to force the Spanish line at Tudela, while at Belchite the bombardment was so heavy that after some caissons had exploded in their midst the Spanish army ran away.

Conclusion

In the later Napoleonic Wars the French artillery perfected the technique of acting in mass at a decisive point after this had been identified in the preliminary outpost battle. This was a tactic which demanded good conditions of weather and terrain, as well as close support from the other arms. It was therefore far from universally successful; but when it did work it could be totally paralysing.

Other nations were quick to see the direction in which the French gunners were heading, and could claim successes of their own with the same techniques. Wellington's guns were massed at Vitoria, albeit accidentally, and Bulow's at Gross Beeren. In particular it was the Russians who always used an enormous weight of artillery, and at Eylau even gave a lesson in technique to the French. Indeed, it was the Russian theorist, Okouneff, who after the wars most clearly laid down the principles of the new tactics. His recipe was that providing you had a superiority in guns of about 25% it would always be possible to establish a battery of 80-120 guns at a decisive point. This battery could then destroy one infantry division every hour, or dismount any enemy artillery as it came up piecemeal. Jomini dismissed this as "exaggerated", but it was very much what Napoleon must often have had in mind.

As the nineteenth century wore on, the problems of deploying an artillery mass at the right moment became increasingly apparent. In the Crimea, 1859, and the American Civil War the administrative and logistical difficulties proved insuperable. By the time of the Franco Prussian War, on the other hand, the increased range of the guns themselves made it unnecessary to "get up close and shoot fast". From that point onwards the massing of artillery fire could be achieved with batteries which were miles apart, out of sight both of each other and of the target. Only with the appearance of the tank could it be said that there was again any sort of "artillery charge" in the Napoleonic sense: only then could the spirit of Sénarmont and Drouot be revived.

Bibliography

The works which I have found most useful are the following, although I have also drawn upon a much wider range of memoirs and private textbooks for individual points. I am particularly grateful to the library staff at Sandhurst, the Staff College, and the French Ministry of War for all their enthusiastic assistance.

Chandler; *The Campaigns of Napoleon* (1967)

Decker; *Traité d'Artillerie* (1825)

Derode; *Nouvelle Rélation de la Bataille de Friedland* (1839)

Du Teil; *De l'Usage de l'Artillerie Nouvelle dans la Guerre de Campagne* (1778)

Escalle; *Des Marches dans les Armées de Napoléon* (1912)

Fallot; *Cours d'Art Militaire* (1837)

Franclet; Manuscript notes on Napoleonic artillery in the library of the French Ministry of War (1926)

Gassendi; *Aide Mémoire à l'Usage des Officiers d'Artillerie en France* (4th. Edn., 1809)

Grois; *Mémoires 1792-1822* (1909)
Head; *French Napoleonic Artillery* (1970)
Hicks; *French Military Weapons 1717-1938* (1964)
Hughes; *Firepower* (1974)
Lachouque and Brown; *The Anatomy of Glory* (2nd. Edn., 1962)
Lauerma; *L'Artillerie de Campagne Francaise Pendant les Guerres de la Révolution* (1956)
Le Bourg; *Essai sur l'Organisation de l'Artillerie et de son Emploi dans la Guerre de Campagne* (1845)
May; *Achievements of Field Artillery* (1893)
Okouneff; *Use of Artillery in the Field* (1856)
Owen; *Modern Artillery* (1873)

Notes

1 Comparative tables of artillery weights are to be found in Hicks' work.

2 The word "limber" is here used as a translation for the pair of wheels joined by an axle upon which the trail of the gun rested while in motion. This was known as the *"avant train"*, but was not a true limber as it was not really an ammunition box. There was a small detachable ammunition case (*"coffret"*) which normally rested upon it, but this could not hold as many rounds as later limbers.

3 *Essai Generale de la Tactique*, new edn., 1803, chapter 5.

4 *Projet d'Ordonnance Provisoire pour l'Artillerie, Contenant l'Ecole et les Manoeuvres d'une Batterie de Campagne* (Paris, 15th. Oct., 1809).